Understanding the Elements of the Periodic Table™

THE 15 LANTHANIDES AND THE 15 ACTINIDES

Kristi Lew

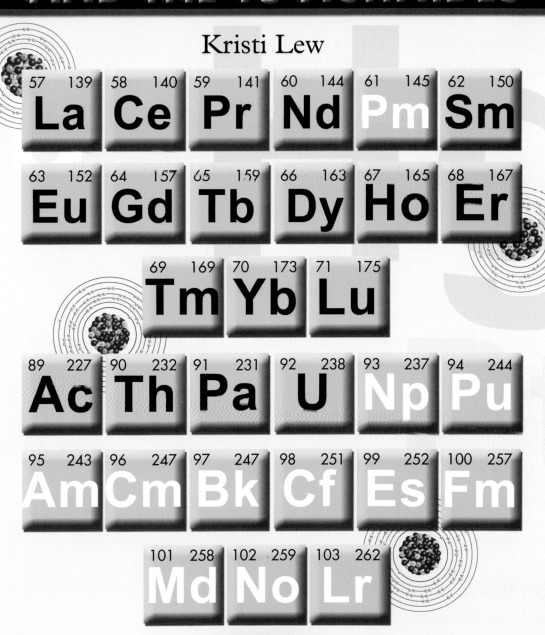

57 139	58 140	59 141	60 144	61 145	62 150
La	Ce	Pr	Nd	Pm	Sm
63 152	64 157	65 159	66 163	67 165	68 167
Eu	Gd	Tb	Dy	Ho	Er
69 169	70 173	71 175			
Tm	Yb	Lu			
89 227	90 232	91 231	92 238	93 237	94 244
Ac	Th	Pa	U	Np	Pu
95 243	96 247	97 247	98 251	99 252	100 257
Am	Cm	Bk	Cf	Es	Fm
101 258	102 259	103 262			
Md	No	Lr			

rosen publishing's
rosen central®

New York

Published in 2010 by The Rosen Publishing Group, Inc.
29 East 21st Street, New York, NY 10010

Library of Congress Cataloging-in-Publication Data

Lew, Kristi.
The 15 lanthanides and the 15 actinides / Kristi Lew.—1st ed.
    p. cm.—(Understanding the elements of the periodic table)
Includes bibliographical references and index.
ISBN 978-1-4358-3557-3 (library binding)
1. Rare earth metals—Popular works. 2. Actinide elements—Popular works. 3. Periodic law—Popular works. I. Title. II. Title: Fifteen lanthanides and the fifteen actinides.
QD172.R2L49 2010
546'.41—dc22

                                                                2009014416

Manufactured in Malaysia

CPSIA Compliance Information: Batch #TW10YA: For Further Information contact Rosen Publishing, New York,
New York at 1-800-237-9932

On the cover: The lanthanides' and actinides' squares on the periodic table of elements

Contents

Introduction

The periodic table of elements is a chart that is used by chemists to organize all the known chemical elements. A chemical element is a substance that is made up of only one type of atom. Chemical elements are the building blocks of all matter. (Matter is anything that has weight and takes up space.) Gold (Au), silver (Ag), and oxygen (O), for example, are all chemical elements. An element cannot be broken down by ordinary physical or chemical means, such as heating it up or passing electricity through it.

One of the first periodic tables was developed in 1869 by a Russian chemist named Dmitry Mendeleyev (1834–1907). To design his periodic table, Mendeleyev arranged all the elements known at that time (about sixty of them) in order of increasing atomic weight. He discovered that the properties of the elements repeated at regular, or periodic, intervals. Therefore, Mendeleyev arranged the elements in rows, or periods. A new row was started whenever the properties of the elements began to repeat. This way, all the elements in the same column, or group, had similar physical properties. These physical properties could include color, hardness, and the temperatures at which they melt and boil. Elements in the same group also behave similarly in chemical reactions.

In 1869, Dmitry Mendeleyev placed all of the known chemical elements into groups with similar properties. This chart became known as the periodic table of the elements. Here is Mendeleyev's first periodic table, published in 1869.

но въ ней, мнѣ кажется, уже ясно выражается примѣнимость вы- ставляемаго мною начала ко всей совокупности элементовъ, пай которыхъ извѣстенъ съ достовѣрностію. На этотъ разъ я и желалъ преимущественно найдти общую систему элементовъ. Вотъ этотъ опытъ:

		Ti=50	Zr=90	?=180.
		V=51	Nb=94	Ta=182.
		Cr=52	Mo=96	W=186.
		Mn=55	Rh=104,4	Pt=197,4
		Fe=56	Ru=104,4	Ir=198.
		Ni=Co=59	Pl=106,6	Os=199.
H=1		Cu=63,4	Ag=108	Hg=200.
	Be=9,4	Mg=24	Zn=65,2	Cd=112
	B=11	Al=27,4	?=68	Ur=116	Au=197?
	C=12	Si=28	?=70	Sn=118
	N=14	P=31	As=75	Sb=122	Bi=210.
	O=16	S=32	Se=79,4	Te=128?
	F=19	Cl=35,5	Br=80	I=127
Li=7	Na=23	K=39	Rb=85,4	Cs=133	Tl=204
		Ca=40	Sr=87,6	Ba=137	Pb=207.
		?=45	Ce=92	
		?Er=56	La=94	
		?Yt=60	Di=95	
		?In=75,6	Th=118?	

While developing his periodic table, Mendeleyev sometimes came to a spot where none of the known elements exhibited the correct properties to fill in the space. When this happened, Mendeleyev left a gap in his chart. He predicted that, someday, scientists would discover elements that would fill in the blanks. He even predicted what properties these elements would have. Some scientists did not believe Mendeleyev, but he was proven correct when the elements gallium (Ga), scandium (Sc), and germanium (Ge) were discovered in 1875, 1879, and 1886, respectively. These elements all possessed the properties that Mendeleyev had predicted. Each fit neatly into blank spaces that he left in his periodic table.

The modern periodic table is no longer arranged by increasing atomic weight. Instead, the elements are ordered according to increasing atomic number. Uranium (U), for example, has an atomic number of 92. It can be found in one of the two rows of elements at the very bottom of the periodic table. These two blocks of elements are labeled the lanthanide series and the actinide series.

The first element in the lanthanide series is lanthanum (La). Lanthanum has an atomic number of 57 and can be found in period 6, group 3 (or IIIB in an older naming system). The lanthanides are sometimes called the rare earth elements. Uranium is an element in the actinide series. The actinide series begins with the element actinium (Ac) at atomic number 89 and continues through lawrencium (Lr), atomic number 103. There are fifteen lanthanide elements and fifteen actinide elements.

Some of the rare earth elements can be very useful in everyday life. Gadolinium (Gd), for example, is used to make compact discs, and terbium (Tb) is found in microprocessor chips used in computers. The actinides uranium and plutonium (Pu) are important, too. These elements are used as fuel sources in nuclear power plants that can produce electricity for cities and towns. Indeed, the lanthanides and actinides can be found in products ranging from movie projectors and color televisions to airplane parts and decorative glass bowls. Without these elements, everyday life would be very different.

Chapter One
What Are the Lanthanides and Actinides?

Of the thirty elements that make up the lanthanides and actinides, a little more than half of them are found naturally in Earth's crust. The others are synthetic and are made in nuclear reactors or particle accelerators.

Naturally Occurring Elements

In 1787, an amateur geologist named Carl Axel Arrhenius (1757–1824) discovered a new mineral near the village of Ytterby, Sweden. He named this new mineral ytterbite. This black stone is made up of many of the lanthanides. This marked the first discovery of the lanthanide elements. And in 1989, the mine where Arrhenius found ytterbite was designated a historical monument.

Seven years after Arrhenius discovered ytterbite, a Finnish chemist named Johan Gadolin (1760–1852) began to study the mineral. Gadolin realized that ytterbite (known today as gadolinite) contained several unknown elements. While Gadolin was never able to isolate any of these elements, one of the lanthanides, gadolinium, would later be named for him to acknowledge his work.

It took nine more years before the first of the lanthanides, cerium (Ce), was identified. It was identified by Martin Heinrich Klaproth (1743–1817),

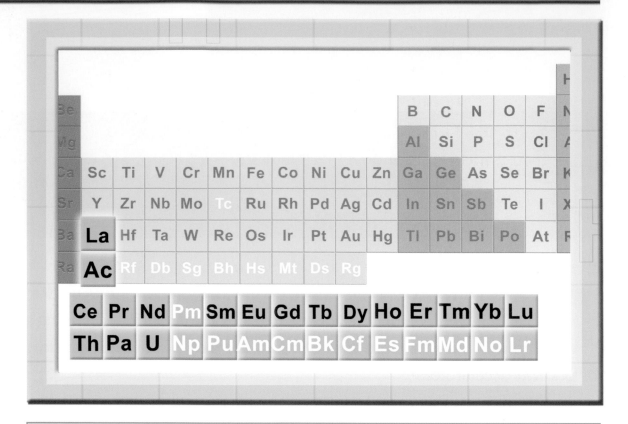

The lanthanides include the fifteen elements from lanthanum to lutetium (Lu). All but one of the lanthanides occur naturally. Most of the actinides, however, are synthetic. They include the fifteen elements from actinium to lawrencium.

a German chemist, as well as Jöns Jacob Berzelius (1779–1848) and Wilhelm von Hisinger (1766–1852), both Swedish chemists. Even though these three scientists are credited for finding cerium, what they actually found was a chemical compound of cerium. A chemical compound is made up of two or more elements that are chemically bonded together. The compound they found was an oxide of cerium (meaning that it contained the element oxygen, as well as cerium). Scientists at that time did not have the technology they needed to isolate the element, but they did name it. The element was named cerium in honor of the first asteroid, Ceres, which had been discovered two years earlier in 1801.

A Swedish chemist named Carl Gustaf Mosander (1797–1858) came to the conclusion that ceria contained at least two oxides. He kept the name ceria for one of them and named the other lanthana, from the Greek word meaning "to lie hidden." It was later determined that lanthana was made up of lanthanum oxide. In 1841, Mosander separated yet another substance from ceria. He named it didymium, from the Greek word for "twin," because he always found it in the same rocks with lanthana. In the next couple of years, Mosander went on to discover two other lanthanides, erbium (Er) in 1842 and terbium in 1843. The American chemists William Frances Hillebrand (1853–1925) and Thomas H. Norton would isolate cerium in 1875.

Forty-four years after Mosander discovered didymium, Carl Auer von Welsbach (1858–1929), a German chemist, separated the substance into two elements: praseodymium (Pr) and neodymium (Nd). He retained the "twin" part of didymium's name and added another Greek word that means "green" to praseodymium's name. He also added the prefix neo, which means "new," to create the name neodymium.

Carl Auer von Welsbach isolated the lanthanides praseodymium and neodymium from didymium.

Rare Earth Elements

Sometimes the lanthanides are called the rare earth elements or rare earth metals. "Earth" is an old chemical word that means "oxide." The lanthanides acquired this name because they are often found chemically bonded to oxygen in nature.

The lanthanides were once considered to be quite rare. Indeed, before 1945, a long and difficult process was required to purify the metals from their compounds. Today, scientists have a better, though still complex, process to isolate the different metals. The lanthanides are no longer considered rare. In fact, even the most elusive of these elements are more common than the metals in the platinum (Pt) group. It turns out that some of these elements, such as lanthanum, cerium, and neodymium, for example, are even more common than lead (Pb).

Making Actinides

Only four actinides occur in nature. Scientists create the rest in a device called a particle accelerator. Particle accelerators are machines that propel tiny particles at extremely high speeds. These particles are smaller than an atom, and they get accelerated to a high-energy state. When these high-energy particles hit an atom, they rearrange the other particles that make up that atom. This process allows new elements to be formed. Neptunium, for example, is created by bombarding uranium atoms with deuterons. Deuterons are subatomic particles that contain one proton and one neutron. Many of the actinides were first made by scientists using the cyclotron particle accelerator at the University of California at Berkeley in the 1940s and 1950s.

Finding Lanthanides

Most of the lanthanides that are used today are isolated from two minerals: monazite and bastnaesite. Large deposits of the mineral monazite are found in the river sands of Brazil and on the beaches of India. The occurrence of monazite sand is widespread, and it can also be found in Florida and North Carolina. Other states, including Wyoming and Colorado, and countries including Norway, Switzerland, and Finland have deposits of monazite stone. Bastnaesite deposits are also common in Norway and Sweden, as well as in areas of Southern California.

Finding Actinides

For the most part, plutonium, neptunium (Np), and the other actinides are artificially created by scientists using nuclear reactors or particle accelerators. Only four of the actinides occur widely in nature: actinium, thorium (Th), protactinium (Pa), and uranium. A few other actinides that were once considered entirely man-made, such as plutonium and neptunium, have since been found in trace amounts in uranium-containing minerals.

Scientists made many actinides by using particle accelerators, such as the cyclotron at the University of California at Berkeley.

The actinide thorium can be found in the mineral monazite along with many of the lanthanides. Scientists estimate that thorium is about as abundant in Earth's crust as lead. The element can also be found in the minerals thorite and thorianite. Large deposits of these minerals exist in New England as well as in other parts of the United States, but they are rarely mined because scientists can get all the thorium that they need from monazite.

Uranium is another naturally occurring actinide. It is most commonly found in a mineral called pitchblende. The German chemist Martin Heinrich Klaproth first discovered uranium oxide in 1789. Like the rare

Uranium oxide, a chemical compound in the mineral uraninite, is radioactive. When X-ray film is placed near a bit of the compound, it is exposed by the radiation, as seen here.

earth elements, uranium was once considered quite scarce. However, scientists now know that this element is more common in Earth's crust than mercury (Hg).

Like uranium, the actinides protactinium and actinium can also be found in the mineral pitchblende. Protactinium is not nearly as abundant as uranium, however. In fact, protactinium makes up only about one part in ten million parts of pitchblende. Protactinium may be the most rare actinide found in nature, but it is not the most rare overall. Even though scientists have known how to make the actinides lawrencium and nobelium (No), for example, for more than fifty years, only tiny amounts of these elements have ever been produced.

Chapter Two
Atomic Structure of the Lanthanides and Actinides

The lanthanides and actinides, as well as all the other elements on the periodic table, are made up of atoms. An atom is the smallest part of an element that still has the properties of that element. Atoms are the building blocks of matter. Matter is anything that has weight and takes up space. A pizza, a car, and the air are all different types of matter, and they are all made up of atoms.

Atomic Structure

Each element on the periodic table is made up of only one particular type of atom. In other words, europium (Eu) is made up of only europium atoms. And europium atoms are different from uranium atoms. Chemists can tell the difference between the atoms of each element based on the number and type of subatomic particles that they consist of.

There are three main types of subatomic particles: protons, electrons, and neutrons. Protons are positively charged subatomic particles that are found in the nucleus, or central core, of an atom. The number of protons in an atom determines what type of atom it is and where it belongs on the periodic table. Each element has a different number of

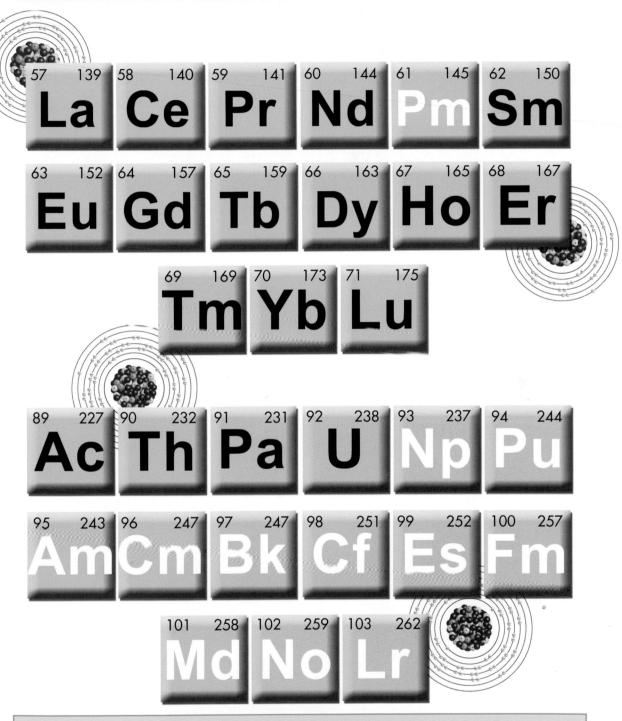

Shown here are the lanthanides' *(top)* and actinides' *(bottom)* squares on the periodic table. Each element has a unique number of protons. The number of protons in an element is equal to the element's atomic number. For example, uranium atoms have 92 protons, and the element's atomic number is 92.

protons. The number of protons in an atom is equal to the atomic number of that element. Europium has an atomic number of 63, which means that it has sixty-three protons. Every uranium atom has ninety-two protons; any atom with more or less than ninety-two protons cannot be an atom of uranium.

Protons have a positive electrical charge, but, overall, atoms are neutral. In other words, they do not have a charge at all. In order for an atom to be electrically neutral, the number of positively charged protons in its nucleus must be balanced by an equal number of negatively charged particles. These negatively charged particles are called electrons. Because an

Many of the lanthanides, including cerium, lanthanum, and ytterbium, can be found in the naturally occurring mineral monazite. The actinide thorium is also found in monazite.

atom of europium has sixty-three protons, it must also have sixty-three electrons in order to be electrically neutral.

Electrons travel around an atom's nucleus in areas called energy levels or shells. The number of energy levels contained in an element's atoms is equal to the element's period number on the periodic table. Europium, for example, is a lanthanide. All the lanthanides fit into the periodic table to the right of lanthanum on period number 6. Therefore, the electrons in europium atoms, and all other lanthanide atoms, have six energy levels. The actinides, on the other hand, belong in period 7. This means that all of their atoms have seven energy levels.

The last of the three main subatomic particles, the neutron, is also found in the nucleus with an atom's protons. Unlike protons and electrons, neutrons do not have a charge. They are electrically neutral.

Atomic Weight

Together, an atom's protons and neutrons make up nearly all the atom's weight. A proton weighs approximately 1.67×10^{-27} kg. Neutrons weigh almost the same. Electrons weigh much, much less. Working with numbers this small, however, can be inconvenient. Therefore, chemists decided that 1.67×10^{-27} kg would be equal to 1 atomic mass unit (amu). This means that the weight of one proton or neutron is equal to 1 amu. The average weight of an atom of each element is listed in the periodic table in atomic mass units. For example, the atomic weight listed on the periodic table for a uranium atom is 238 amu. Because all uranium atoms must have 92 protons that weigh 92 amu, the other 146 amu must come from neutrons. Nearly all uranium atoms have 146 neutrons.

However, not all uranium atoms have 146 neutrons. There are other less common types of uranium atoms. One type has an atomic weight of

Radioactivity

The number of neutrons in an atom's nucleus helps determine its stability. Protons in the nucleus have the same electric charge, and like charges repel one another. This means that each proton is trying to get away from the other protons. Neutrons help isolate protons so they don't repel each other. If an atom's nucleus does not have enough neutrons, the protons can push each other apart. This causes the nucleus to break down, or decay, over time. When a nucleus decays, it releases energy in the form of rays or particles. This energy is called radiation. Elements that undergo the process of nuclear decay are called radioactive. All of the actinides and one of the lanthanides (promethium) are radioactive elements.

The time that it takes for half the atoms in a radioactive sample to decay is called the half-life. The isotopes of some radioactive elements have very long half-lives. The most common type of uranium atom, for example, has a half-life of nearly 4.5 billion years. One thorium isotope lasts even longer—approximately fourteen billion years. Other isotopes have much shorter half-lives. One isotope of nobelium, for example, has a half-life of only 2.3 seconds.

235 amu, and the other type has a weight of 234 amu. While all these types of uranium atoms have ninety-two protons, they do not all have the same number of neutrons. This explains the difference in their weights. Forms of the same element that have different numbers of neutrons and, therefore, different atomic weights are called isotopes.

All of the actinides, including uranium, are radioactive elements. The radiation given off by minerals that contain uranium can be harmful to human health.

Properties of the Lanthanides and Actinides

On the periodic table, elements with similar properties are listed in the same group. The reason elements in the same group tend to have similar properties is due to the arrangement of their electrons. The elements lithium (Li) and sodium (Na), for example, are in the same group (group 1 or IA) on the periodic table. These elements also react in similar ways during chemical reactions. Lithium has an atomic number of 3. This means that all lithium atoms have three protons. In order for them to be electrically neutral, they must also have three electrons. Lithium is in period 2. Therefore, its three electrons are arranged in two energy levels. Two of those electrons can fit in the first energy level. The remaining electron goes in the second energy level. This electron in the outermost, or highest, energy level is the electron that is involved in chemical reactions. The outermost electrons in an atom are called valence electrons. Lithium has one electron in its highest energy level, so it has one valence electron.

Sodium has eleven electrons arranged in three energy levels—two in the first energy level, eight in the second, and one in the third. Like lithium, sodium has one electron in its highest energy level, and it has one valence electron. Because lithium and sodium have the same number of valence electrons, they have similar properties.

The lanthanides and actinides are not in the same columns on the periodic table. However, the lanthanides all have remarkably similar properties. This is because their atoms all have the same number of valence electrons, namely two in the sixth level. The different lanthanide elements have different numbers of electrons in their atoms, but the differences are in the fourth energy level, not in the outermost level. The actinides also all tend to react in the same ways. This is because they all have two valence electrons in the seventh energy level. The atoms of the actinide elements have different numbers of electrons, but the differences are in the fifth energy level, which does not affect their properties.

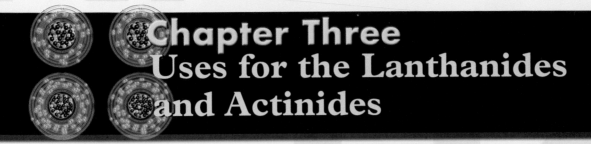
Because the rare earth metals have similar properties, many of them can be used in the same applications. The lanthanides are used in lamps, motion picture projectors, lasers, magnets, and phosphors. These elements are also useful for making different colored decorative glass and ceramic products.

Alloys

An alloy is a mixture of two or more elements, at least one of which is a metal. Alloys are used because they have different properties than the pure metals that are mixed together.

Many of the lanthanides are alloyed with other metals. Thorium and praseodymium, for example, can be alloyed with magnesium to make airplane engine parts that remain strong even at high temperatures. Gadolinium can improve a metal's resistance to corrosion, and it is often added to iron. Alloys of praseodymium and neodymium are used to make magnets, and erbium is alloyed with the metal vanadium (V) to make it softer and easier to shape.

One of the oldest uses for some of the lanthanides is in a material called misch metal. This metal is an alloy of 50 percent cerium, 25 percent

The lanthanides are often alloyed with other metals. These alloys have many applications and are often used to make strong, heat-resistant airplane engine parts.

lanthanum, 15 percent neodymium, and 10 percent other rare earth elements and iron. When misch metal is struck, it gives off sparks. It is often used to make the flints in lighters.

Nuclear Power

Most of the actinides are synthetically made, have very short half-lives, and are radioactive. As a result, the uses for these pure elements are limited. However, some of the actinides can be used to make electrical power. These elements go through nuclear fission reactions as they decay. This process releases a lot of energy, mainly in the form of heat.

Two workers at a nuclear power plant stand near a pool containing nuclear fuel. Uranium is the most common fuel used in today's nuclear power reactors.

In fact, scientists believe that most of Earth's internal heat comes from the radioactive decay of the actinides thorium and uranium. Both of these elements, along with plutonium, are also sources of nuclear energy. Scientists estimate that there is more power available in the thorium naturally found in Earth's crust than in uranium and fossil fuels combined.

For now, uranium is the primary element used in nuclear reactors. Inside a nuclear reactor, uranium atoms are bombarded with neutrons. These collisions cause the uranium atoms to break apart into atoms of different elements such as krypton (Kr) and barium (Ba), or strontium (Sr) and xenon (Xe). This reaction is known as fission. Fission also releases large amounts of energy. This energy can be captured and used to make electricity. The fission reaction also produces additional neutrons. These neutrons go on to hit other uranium atoms in the reactor, continuing the fission reactions. This cascade of fission reactions is called a chain reaction.

Common Alloys

The lanthanides are not the only elements that are used to make alloys. Brass, for example, is a mixture of copper (Cu) and zinc (Zn). Brass is more malleable than either copper or zinc alone. Malleability is a physical property of a metal that describes how easily it can be hammered into different shapes without breaking. Bronze and stainless steel are other common alloys. Bronze is a mixture of copper and tin (Sn). It has a lower melting point than pure copper, so it is easier to melt and pour into molds to make metal parts. Stainless steel is an alloy of chromium (Cr) and iron (Fe). While pure iron will rust, stainless steel will not. It is used in products to make them resistant to corrosion.

In a nuclear power station, the heat generated during fission reactions is captured and used to heat water to a boil. The boiling water produces steam that turns large blades in a mechanism called a turbine. The turbine's fan runs a generator that creates electricity for industries and for people's homes, offices, and schools.

Nuclear power is generally considered to be a cleaner way to make electricity than burning coal. It does not produce soot or chemical pollution, and it does not release harmful gases into the atmosphere. However, nuclear power is not without its problems. The transport of nuclear fuel, for example, can pose a risk for people living near highways or train

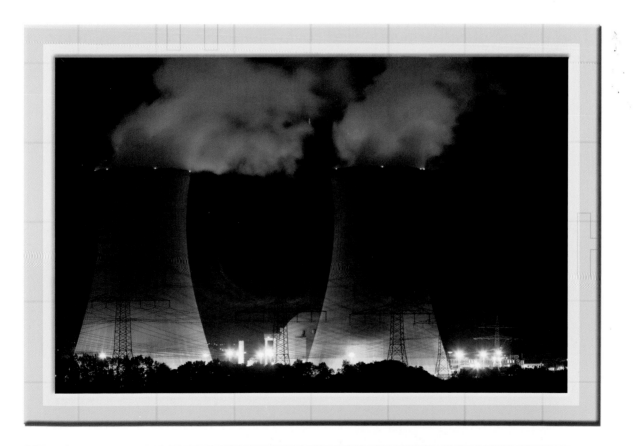

This German nuclear power plant is one of many in the country. Nuclear power can provide clean energy. Despite this, there are many safety concerns surrounding nuclear power and the disposal of spent nuclear fuel.

tracks that carry the fuel. In the past, accidents have occurred at nuclear power plants. While this is rare, it is still a concern.

Storing the spent fuel from nuclear power stations is another safety consideration. Spent nuclear fuel can no longer produce the energy needed to create electricity and, therefore, it must be removed from the nuclear reactor. However, that does not mean that this fuel is safe. It is even more radioactive than the original fuel, and it needs to be stored where the radiation cannot harm people. Eventually, the spent fuel will decay into stable elements, but this can take tens of thousands of years.

Scientists also use some of the actinides to produce the neutrons needed to make other actinides. Actinium and californium, for example, are good neutron makers. Just 1 microgram (35 millionths of an ounce) of an isotope of californium called californium-252 can emit as many as 170 million neutrons per minute.

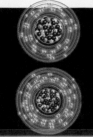

Chapter Four
The Lanthanides and Actinides in Compounds

When the lanthanides and actinides chemically react with other elements to make chemical compounds, they form a type of chemical bond called an ionic bond. An ionic bond is formed when one atom loses one or more electrons and another atom accepts them.

Ions

The lanthanides and actinides are metals. When metal atoms meet nonmetal atoms, the metal atoms give up one or more electrons to the nonmetal atoms. When a metal atom loses one or more electrons, it has fewer electrons than protons. Therefore, it is no longer a neutral atom. The extra proton or protons gives the particle a positive charge.

Nonmetal atoms like oxygen, chlorine, and fluorine accept the electrons that the metals give up. The extra electron or electrons cause the nonmetal atoms to have a negative charge.

Atoms that have a positive charge after losing or gaining electrons are called ions. A positive metal ion and a negative nonmetal ion have opposite charges. Opposite charges attract each other, so the positive metal ion is attracted to the negative nonmetal ion. This attraction forms an ionic

Several of the lanthanide oxides are used to make the lenses for cameras, microscopes, and telescopes.

bond between ions of opposite charges. Ionic compounds can also be called salts. The salts of the lanthanides and actinides have many uses.

Glassmaking

When light passes through a lens, the lens changes the direction of the light. High-index glass changes the direction more than regular glass. This allows lenses made from high-index glass to be smaller and lighter than those made from regular glass. Oxides of the rare earth metals are used to make some of the highest-index glass available.

Glass with other useful properties can also be made from lanthanide compounds. The elements praseodymium and neodymium are often used in certain types of goggles. The glass in these goggles, called didymium glass, blocks bright yellow light that is emitted when a welder's torch heats metal or a glassblower's torch heats glass. When samarium oxide is added to glass, the glass can shield a worker from infrared (IR) light. Infrared light is felt as heat.

The addition of rare earth elements to glass can also give it other beneficial properties. When lanthanum oxide is added to glass, it can help the glass resist the corrosive effects of very strong chemicals. The salts of

What's the Use?

Many of the lanthanides and actinides have no practical purpose at all. Sometimes this is because the elements are too expensive. This expense comes from the fact that they are difficult to extract and purify. Some actinides don't have many practical applications because only very small amounts of them have been made. Also, they are often dangerously radioactive.

Some of the lanthanides and actinides have no use today, but they have unique properties that scientists think may come in handy in the future. Dysprosium (Dy), for example, has a very high melting point and, therefore, may be used one day in nuclear power plants to control heat distribution. Very few uses have been found for holmium (Ho) so far, but the element does have some interesting magnetic properties that scientists may be able to exploit in the future. Gadolinium is another lanthanide with some unique magnetic properties. Today, the element is used as a component in compact discs. In the future, scientists believe they might be able to use the element to make magnets that could sense hot and cold.

other rare earth elements are used to color decorative glass, too. For example, erbium oxide gives glass a pinkish tint.

Catalysts

Some lanthanide and actinide compounds are used as catalysts in industrial processes. A catalyst is a substance that can start or speed up a

Some compounds of the lanthanides and actinides help catalyze industrial processes. Oil refineries, such as the one pictured here, use oxides of thorium and cerium to turn petroleum into useful products.

chemical reaction without being used up in the chemical reaction itself. Compounds of cerium, samarium (Sm), lutetium, and thorium are all used to catalyze different processes. Thorium oxide, for example, can be used as a catalyst in the production of two other very important chemicals: nitric acid and sulfuric acid. Both thorium and cerium oxide are used as catalysts in the refining of petroleum (turning it into diesel fuel, kerosene, or gasoline). Cerium oxide is also used as a catalyst in the coating found inside self-cleaning ovens.

Glowing Bright

Rare earth elements can be used to produce very bright light. A type of lamp called a carbon arc lamp was once used in lighthouses and searchlights. It was also used by the motion picture industry as studio lighting and for projector lights. When electricity was passed through the electrodes of these lamps, they would heat up and give off a brilliant light. The electrodes were made of carbon and small amounts of other elements. The color of the light depended on which trace element or elements were added to the carbon. Oxides and other compounds of lanthanum, cerium, praseodymium, or samarium were often used for these lamps.

Another place where lanthanides produce extremely bright light is in gas lamps. Before the invention of electric lights, gas lamps were used to light private homes and city streets. Today, gas lamps are used by campers. The flame in gas lamps is surrounded by a piece of gauze known as the mantle. The mantle is dipped in cerium oxide. When the mantle heats up, it gives off a very bright light.

Compounds of the lanthanides are also used in many color computer monitors and television sets with cathode ray tube (CRT) screens. These compounds are used to make phosphors. A phosphor is a substance

A phosphor painted onto the top of this cathode ray tube glows green when hit with a stream of electrons. The lanthanide compound terbium oxide is sometimes used as an activator for green phosphors.

that gives off light when it is hit with electrons. CRT screens have phosphors painted on the back of the screen inside the tube. (Liquid crystal display [LCD] and plasma screens use a different technology.) The color of light produced by the phosphor depends on which elements are in it. CRT screens are coated with tiny dots of three different phosphors: a red one, a green one, and a blue one. The different colors seen on the monitor or television are made by mixing these three colors. Compounds of yttrium and the lanthanides gadolinium and europium are used to make red phosphors. Terbium oxides may be used as activators for green phosphors, and lanthanum phosphate can be used to make blue phosphors.

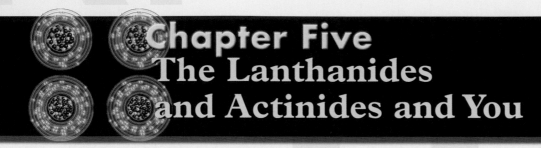

Chapter Five
The Lanthanides
and Actinides and You

While some of the lanthanides are used to make computers, televisions, and compact discs, these are not the only everyday products that we use that contain these elements. Lanthanides are also used in hospitals to help doctors diagnose and treat diseases. Actinides are used in the creation of electricity and smoke detectors. In these cases, these radioactive elements are very useful. However, they can also be quite dangerous if they are not handled properly.

Medicine

Because the element gadolinium has some unique magnetic properties, it can be used in a test called a magnetic resonance imaging (MRI) scan. An MRI scan uses radio waves and a magnetic field to create a clear and detailed picture of the inside of the body. To get the best pictures, substances called contrast agents are sometimes swallowed or injected before the test is given. In many MRI tests, gadolinium is used as the contrast agent. Gadolinium builds up in abnormal tissue. This makes the abnormal tissue (such as a tumor, for example) show up as a bright spot on the picture.

Gadolinium is used as a contrast agent in MRI scans. It makes abnormal tissue, such as a tumor, stand out from the tissue around it. In this MRI scan of a person's head, a tumor appears as a green blotch amid healthy brain tissue.

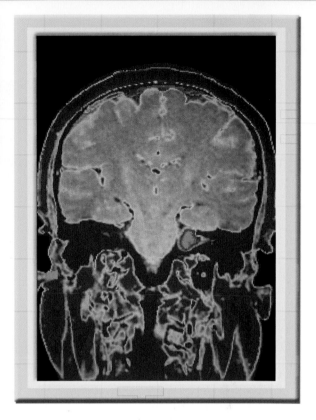

Gadolinium is not the only lanthanide that is used by doctors. A compound of lanthanum, called lanthanum carbonate, has been approved as a medication for people who have kidney disease. When someone has kidney disease, certain chemicals that are normally filtered out by the kidneys, such as sodium, potassium, and phosphate compounds, can build up in the body. Too much phosphate in the blood can lead to bone disease, cause severe itchiness, and possibly damage the patient's arteries. Lanthanum carbonate is used to absorb excess phosphate in the blood so it does not cause these problems.

Radioactivity and the Human Body

Uranium and plutonium are crucial to the nuclear power industry. However, the radiation that is emitted by these elements—either in their natural state, in a nuclear weapon, as spent nuclear fuel, or during a nuclear power plant accident—can present a significant risk to human health.

During nuclear decay reactions, radioactive atoms change and become more stable elements. To become a more stable element, the

Colorful Elements

Many of the lanthanides make our world just a little prettier. They do this by adding color to glass, enamels, and glazes that are used to make decorative objects such as vases, lamps, and plates. Erbium oxide, for example, gives glass and glazes a delicate pink color. Salts of praseodymium are used to color it yellow. Neodymium compounds can add the colors violet, red, or gray, depending on which compound is used. Sometimes, however, glass with no color is more desirable. In that case, cerium and neodymium compounds can be used to remove unwanted color from glass.

Vaseline glass is a pretty yellow-green color. But because it contains a radioactive compound of uranium, it must be handled with care.

People have also used one of the actinides to add color to decorative objects. "Vaseline" glass (also known as uranium glass) is a yellow-green glass that turns a florescent green color under black light. (Black light gives off UV light.) Vaseline glass is made using a salt of uranium. However, pure uranium and its compounds are radioactive. Therefore, these compounds are no longer used for this purpose, and any items made with them should be handled carefully.

radioactive element needs to change the ratio of protons to neutrons in its nucleus. One way to do this is to eject some subatomic particles from the nucleus.

The actinide americium (Am), for example, decays by emitting alpha particles. An alpha particle is a type of radiation made up of two protons and two neutrons emitted from the nucleus of a radioactive element. Of all the different types of radiation, alpha particles are the largest. They also contain the least amount of energy, and they are among the safest. In fact, alpha particles do not have enough energy to penetrate a piece of notebook paper or the outer layer of skin. So even though americium is a radioactive element, it is often used in some types of smoke detectors found in people's homes. The alpha particles emitted by the americium are used to help create a small electric current in the smoke detector. When smoke enters the detector, it disrupts this electric current and causes the smoke alarm to sound. As long as the smoke detector is not damaged, the alpha particles emitted by the americium cannot escape and the americium does not pose a health risk to the people who live in the house with it.

Other types of radioactive nuclei become more stable by ejecting beta particles. Essentially, a beta particle is an electron that originated in the nucleus of a radioactive element. Elements that emit this type of radiation change the ratio of protons to neutrons in their nucleus by converting a neutron into a proton and an electron. The proton stays in the nucleus, but the electron is ejected. With their high energy, beta particles can harm the human body by breaking the bonds between atoms and forming ions. Thorium uses a combination of alpha particle and beta particle decay to become a stable isotope of lead.

Along with alpha and beta particles, radioactive elements can also emit a type of high-energy radiation called gamma rays. Gamma rays are energy waves that travel at the speed of light. They can easily penetrate

Many smoke detectors contain the radioactive actinide americium. Radiation can be harmful to humans, but as long as it is safely contained, it should pose no serious danger.

the human body and damage the body's cells. If cells are damaged in a certain way, they can grow out of control. Cells growing out of control can cause cancer to grow within the body.

While radiation can be dangerous, without the lanthanides and actinides we would not be able to make many of the objects that we use on a daily basis: lightweight and strong airplane parts, decorative glass objects, professional-grade camera lenses, important scientific instruments, CRT monitors and television screens, and inexpensive smoke detectors. Some people's health would suffer without the medication and medical tests that are necessary to help them feel better. And some uses for these elements have yet to be discovered. Yes, some of the lanthanides and actinides can be harmful. But if they are handled with care, they can be extremely useful to humankind.

The Periodic Table of Elements

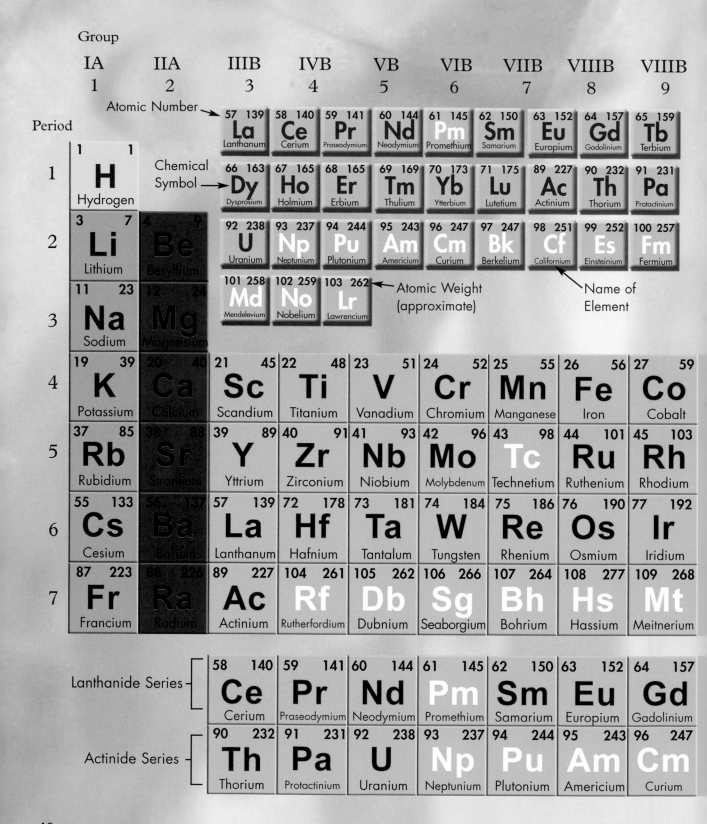

Group

| IA 1 | IIA 2 | IIIB 3 | IVB 4 | VB 5 | VIB 6 | VIIB 7 | VIIIB 8 | VIIIB 9 |

Atomic Number →

Period

| 57 139 La Lanthanum | 58 140 Ce Cerium | 59 141 Pr Praseodymium | 60 144 Nd Neodymium | 61 145 Pm Promethium | 62 150 Sm Samarium | 63 152 Eu Europium | 64 157 Gd Gadolinium | 65 159 Tb Terbium |

Chemical Symbol →

| 66 163 Dy Dysprosium | 67 165 Ho Holmium | 68 165 Er Erbium | 69 169 Tm Thulium | 70 173 Yb Ytterbium | 71 175 Lu Lutetium | 89 227 Ac Actinium | 90 232 Th Thorium | 91 231 Pa Protactinium |

| 92 238 U Uranium | 93 237 Np Neptunium | 94 244 Pu Plutonium | 95 243 Am Americium | 96 247 Cm Curium | 97 247 Bk Berkelium | 98 251 Cf Californium | 99 252 Es Einsteinium | 100 257 Fm Fermium |

Name of Element

| 101 258 Md Mendelevium | 102 259 No Nobelium | 103 262 Lr Lawrencium | Atomic Weight (approximate) |

1 1 H Hydrogen								
3 7 Li Lithium	4 9 Be Beryllium							
11 23 Na Sodium	12 24 Mg Magnesium							
19 39 K Potassium	20 40 Ca Calcium	21 45 Sc Scandium	22 48 Ti Titanium	23 51 V Vanadium	24 52 Cr Chromium	25 55 Mn Manganese	26 56 Fe Iron	27 59 Co Cobalt
37 85 Rb Rubidium	38 88 Sr Strontium	39 89 Y Yttrium	40 91 Zr Zirconium	41 93 Nb Niobium	42 96 Mo Molybdenum	43 98 Tc Technetium	44 101 Ru Ruthenium	45 103 Rh Rhodium
55 133 Cs Cesium	56 137 Ba Barium	57 139 La Lanthanum	72 178 Hf Hafnium	73 181 Ta Tantalum	74 184 W Tungsten	75 186 Re Rhenium	76 190 Os Osmium	77 192 Ir Iridium
87 223 Fr Francium	88 226 Ra Radium	89 227 Ac Actinium	104 261 Rf Rutherfordium	105 262 Db Dubnium	106 266 Sg Seaborgium	107 264 Bh Bohrium	108 277 Hs Hassium	109 268 Mt Meitnerium

Period numbers (left): 1, 2, 3, 4, 5, 6, 7

Lanthanide Series

| 58 140 Ce Cerium | 59 141 Pr Praseodymium | 60 144 Nd Neodymium | 61 145 Pm Promethium | 62 150 Sm Samarium | 63 152 Eu Europium | 64 157 Gd Gadolinium |

Actinide Series

| 90 232 Th Thorium | 91 231 Pa Protactinium | 92 238 U Uranium | 93 237 Np Neptunium | 94 244 Pu Plutonium | 95 243 Am Americium | 96 247 Cm Curium |

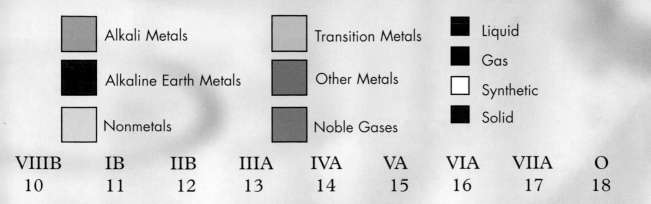

					2 4 **He** Helium

5 11 **B** Boron	6 12 **C** Carbon	7 14 **N** Nitrogen	8 16 **O** Oxygen	9 19 **F** Fluorine	10 20 **Ne** Neon
13 27 **Al** Aluminum	14 28 **Si** Silicon	15 31 **P** Phosphorus	16 32 **S** Sulfur	17 35 **Cl** Chlorine	18 40 **Ar** Argon

28 59 **Ni** Nickel	29 64 **Cu** Copper	30 65 **Zn** Zinc	31 70 **Ga** Gallium	32 73 **Ge** Germanium	33 75 **As** Arsenic	34 79 **Se** Selenium	35 80 **Br** Bromine	36 84 **Kr** Krypton
46 106 **Pd** Palladium	47 108 **Ag** Silver	48 112 **Cd** Cadmium	49 115 **In** Indium	50 119 **Sn** Tin	51 122 **Sb** Antimony	52 128 **Te** Tellurium	53 127 **I** Iodine	54 131 **Xe** Xenon
78 195 **Pt** Platinum	79 197 **Au** Gold	80 201 **Hg** Mercury	81 204 **Tl** Thallium	82 207 **Pb** Lead	83 209 **Bi** Bismuth	84 209 **Po** Polonium	85 210 **At** Astatine	86 222 **Rn** Radon
110 271 **Ds** Darmstadtium	111 272 **Rg** Roentgenium							

65 159 **Tb** Terbium	66 163 **Dy** Dysprosium	67 165 **Ho** Holmium	68 167 **Er** Erbium	69 169 **Tm** Thulium	70 173 **Yb** Ytterbium	71 175 **Lu** Lutetium
97 247 **Bk** Berkelium	98 251 **Cf** Californium	99 252 **Es** Einsteinium	100 257 **Fm** Fermium	101 258 **Md** Mendelevium	102 259 **No** Nobelium	103 262 **Lr** Lawrencium

alloy A mixture of two or more elements, at least one of which is a metal.

atom The smallest part of an element that still has properties of that element; the building block of all matter.

atomic number The number of protons in an atom of an element.

catalyst A substance that can start or speed up a chemical reaction without being used up in the reaction.

compound A substance made up of two or more elements that are chemically bonded together.

electron A negatively charged subatomic particle found traveling around the nucleus of an atom in energy levels, or shells.

element A chemical substance that is made up of only one type of atom.

group A column of elements on the periodic table.

half-life The amount of time that it takes for half the atoms in a radioactive sample to decay.

ion A charged particle that is formed when an atom loses or gains electrons.

matter Anything that has weight and takes up space.

neutron An uncharged subatomic particle found inside the nucleus of an atom.

nucleus The central core of an atom.

period A row of elements on the periodic table.

proton A positively charged subatomic particle found inside the nucleus of an atom.

radiation Energy released in the form of rays or particles.

valence electrons Electrons in the outermost or highest energy level of an atom and that participate in chemical bonding.

American Chemical Society
1155 Sixteenth Street NW
Washington, DC 20036
(800) 227-5558
Web site: http://portal.acs.org/portal/acs/corg/content
The world's largest scientific society, the American Chemical Society has
    been in existence for more than one hundred years.

Argonne National Laboratory
Division of Educational Programs
9700 South Cass Avenue, Building 223
Argonne, IL 60439
(630) 252-4114
Web site: http://www.anl.gov
The Argonne National Laboratory is one of the U.S. Department of
    Energy's largest research centers and has several particle accelera-
    tors that it uses for primary research.

Atomic Energy of Canada Limited
2251 Speakman Drive
Mississauga, ON L5K 1B2
Canada
(866) 513-2325
Web site: http://www.aecl.ca
Atomic Energy of Canada Limited provides information on the benefits
    and safety of using nuclear energy, how nuclear reactors work, and
    facts about radiation.

Los Alamos National Laboratory
1619 Central Avenue, MS A117
Los Alamos, NM 87545
(888) 841-8256
Web site: http://www.lanl.gov
Los Alamos National Laboratory is a national security research laboratory that investigates problems in bioscience, chemistry, computer science, environmental science, and many other scientific disciplines. It hosts the Los Alamos Space Science Outreach Summer Institute for teacher and student internships each summer.

U.S. Environmental Protection Agency
Ariel Rios Building
1200 Pennsylvania Avenue NW
Washington, DC 20460
Web site: http://www.epa.gov
The Environmental Protection Agency provides information about the positive and negative aspects of nuclear energy.

Web Sites

Due to the changing nature of Internet links, Rosen Publishing has developed an online list of Web sites related to the subject of this book. This site is updated regularly. Please use this link to access the list:

http://www.rosenlinks.com/uept/lant

For Further Reading

Barber, Ian. *Sorting the Elements: The Periodic Table at Work*. Vero Beach, FL: Rourke Publishing, 2008.

Brent, Lynnette. *Elements and Compounds*. New York, NY: Crabtree Publishing Company, 2008.

Jerome, Kate. *Atomic Universe: The Quest to Discover Radioactivity*. Washington, DC: National Geographic Society, 2006.

Kirkland, Kyle. *Atoms and Materials*. New York, NY: Facts On File, Inc., 2007.

Manning, Phillip. *Essential Chemistry: Atoms, Molecules, and Compounds*. New York, NY: Facts On File, Inc., 2007.

McLeish, Ewan. *The Pros and Cons of Nuclear Power*. New York, NY: Rosen Publishing Group, 2007.

Newmark, Ann, and Laura Buller. *Chemistry*. New York, NY: DK Children, 2005.

Parker, Victoria. *Chernobyl 1986: An Explosion at a Nuclear Power Station*. Chicago, IL: Heinemann, 2006.

Raum, Elizabeth. *Nuclear Energy*. Chicago, IL: Heinemann, 2008.

Roza, Greg. *Plutonium* (Understanding the Elements of the Periodic Table). New York, NY: Rosen Publishing Group, 2008.

Bibliography

Amethyst Galleries' Mineral Gallery. "The Mineral Monazite." Retrieved January 31, 2009 (http://www.galleries.com/minerals/phosphat/ monazite/monazite.htm).

Brain, Marshall. "How Smoke Detectors Work." HowStuffWorks.com. Retrieved January 31, 2009 (http://home.howstuffworks.com/ smoke3.htm).

Brain, Marshall, and Robert Lamb. "Pros and Cons of Nuclear Power Plants." HowStuffWorks.com. Retrieved January 31, 2009 (http://science.howstuffworks.com/nuclear-power5.htm).

EdREN. "Diet in Chronic Renal Failure and CKD." Retrieved January 31, 2009 (http://renux.dmed.ed.ac.uk/EdREN/EdRenINFObits/ Diet_CRF.html).

Fray, Derek. "Chemical Engineering: Separating Rare Earth Elements." *Science*, September 29, 2000. Retrieved January 31, 2009 (http:// www.sciencemag.org/cgi/content/summary/289/5488/2295).

Los Alamos National Laboratory. "A Periodic Table of the Elements at Los Alamos National Laboratory." Retrieved January 31, 2009 (http://www.periodic.lanl.gov/default.htm).

New Materials Asia. "Praseodymium Alloy Suitable for Multi-pole Magnets." FindArticles.com, February 2006. Retrieved January 2009 (http://findarticles.com/p/articles/mi_hb5799/is_/ ai_n29276697).

TechFAQ. "How Does Color Television Work?" Retrieved January 2009 (http://www.tech-faq.com/how-does-color-television-work.shtml).

University of Colorado at Boulder. "Rare Earths." Retrieved January 31, 2009 (http://www.colorado.edu/physics/2000/periodic_table/ rare_earths.html).

Index

About the Author

Kristi Lew is the author of more than twenty-five science books for teachers and students. Fascinated by science at an early age, Lew studied biochemistry and genetics in college. After spending years in genetics laboratories and high school classrooms, she now specializes in writing about science, health, and the environment.

Photo Credits

Cover, pp. 1, 8, 15, 40–41 by Tahara Anderson; p. 5 Library of Congress Prints and Photographs Division; p. 9 © SPL/Photo Researchers, Inc.; p. 11 © Ed Young/Corbis; p. 12 © Joel Arem/Photo Researchers, Inc.; pp. 16, 19 © Biophoto Associates/Photo Researchers, Inc.; p. 22 © Jurnasyanto Sukarno/epa/Corbis; p. 23 © Kay Nietfeld/dpa/Corbis; p. 25 Ralph Orlowski/Getty Images; p. 28 Shutterstock.com; p. 30 Christopher Furlong/Getty Images; p. 32 © Andrew Lambert Photography/Photo Researchers, Inc.; p. 35 © Mehau Kulyk/Photo Researchers, Inc.; p. 36 © Michael Boys/Corbis; p. 38 © Chris Priest/Photo Researchers, Inc.

Designer: Tahara Anderson; Photo Researcher: Cindy Reiman